Busy Ant Maths

Problem Solving and Reasoning Pupil Book 3

Peter Clarke

William Collins' dream of knowledge for all began with the publication of his first book in 1819. A self-educated mill worker, he not only enriched millions of lives, but also founded a flourishing publishing house. Today, staying true to this spirit, Collins books are packed with inspiration, innovation and practical expertise. They place you at the centre of a world of possibility and give you exactly what you need to explore it.

Collins. Freedom to teach.

Published by Collins
An imprint of HarperCollins*Publishers*
The News Building
1 London Bridge Street
London
SE1 9GF

Browse the complete Collins catalogue at
www.collins.co.uk

© HarperCollins*Publishers* Limited 2018

10 9 8 7 6 5 4 3 2 1

ISBN 978-0-00-826056-9

The author wishes to thank Brian Molyneaux for his valuable contribution to this publication.

British Library Cataloguing in Publication Data
A Catalogue record for this publication is available from the British Library

Author: Peter Clarke
Publishing manager: Fiona McGlade
Editor: Amy Wright
Copyeditor: Catherine Dakin
Proofreader: Tanya Solomons
Answer checker: Steven Matchett
Cover designer: Amparo Barrera
Internal designer: 2hoots Publishing Services
Typesetter: Ken Vail Graphic Design
Illustrator: Eva Sassin
Production controller: Sarah Burke
Printed and bound by Martins the Printers

ntents

‹ing mathematical problems

Reasoning mathematically

...ng and applying mathematics in real-world contexts

How to use this book

Aims

This book aims to provide teachers with a resource that enables pupils to:

- develop mathematical problem solving and thinking skills
- reason and communicate mathematically
- use and apply mathematics to solve problems.

The three different types of mathematical problem solving challenge

This book consists of three different types of mathematical problem solving challenge:

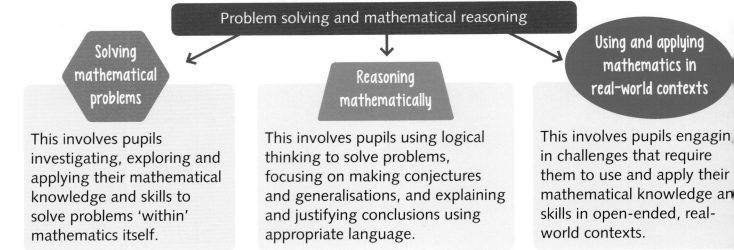

Problem solving and mathematical reasoning

Solving mathematical problems

This involves pupils investigating, exploring and applying their mathematical knowledge and skills to solve problems 'within' mathematics itself.

Reasoning mathematically

This involves pupils using logical thinking to solve problems, focusing on making conjectures and generalisations, and explaining and justifying conclusions using appropriate language.

Using and applying mathematics in real-world contexts

This involves pupils engaging in challenges that require them to use and apply their mathematical knowledge and skills in open-ended, real-world contexts.

This book is intended as a 'dip-in' resource, where teachers choose which of the three different types of challenge they wish pupils to undertake. A challenge may form the basis of part of or an entire mathematics lesson. The challenges can also be used in a similar way to the weekly bank of 'Learning activities' found in the *Busy Ant Maths* Teacher's Guide. It is recommended that pupils have equal experience of all three types of challenge during the course of a term.

The 'Solving mathematical problems' and 'Reasoning mathematically' challenges are organised under the different topics (domains) of the 2014 National Curriculum for Mathematics. This is to make it easier for teachers to choose a challenge that corresponds to the topic they are currently teaching, thereby providing an opportunity for pupils to practise their pure mathematical knowledge and skills in a problem solving context. These challenges are designed to be completed during the course of a lesson.

The 'Using and applying mathematics in real-world contexts' challenges have not been organised by topic. The very nature of this type of challenge means that pupils are drawing on their mathematical knowledge and skills from several topics in order to investigate challenges arising from the real world. In many cases these challenges will require pupils to work on them for an extended period, such as over the course of several lessons, a week or during a particular unit of work. An indication of which topics each of these challenges covers can be found on page 5.

iefing

with other similar teaching and learning resources, pupils will engage more fully with each challenge he teacher introduces and discusses the challenge with the pupils. This includes reading through the allenge with the pupils, checking prerequisites for learning, ensuring understanding and clarifying any sconceptions.

rking collaboratively

e challenges can be undertaken by individuals, pairs or groups of pupils, however they will be enhanced eatly if pupils are able to work together in pairs or groups. By working collaboratively, pupils are more ely to develop their problem solving, communicating and reasoning skills.

u will need

of the challenges require pupils to use pencil and paper. Giving pupils a large sheet of paper, such A3 or A2, allows them to feel free to work out the results and record their thinking in ways that are propriate to them. It also enables pupils to work together better in pairs or as a group, and provides m with an excellent prompt to use when sharing and discussing their work with others.

important problem solving skill is to be able to identify not only the mathematics, but also what ources to use. For this reason many of the challenges do not name the specific resources that are ded.

aracters

e characters on the right are the teacher d the four children who appear in some the challenges in this book.

Mr Johnson

Alexander

Emily

Kate

Osaru

ink about ...

challenges include prompting questions that provide both a springboard and a means of assisting pupils accessing and working through the challenge.

at if?

e challenges also include an extension or variation that allows pupils to think more deeply about the allenge and to further develop their thinking skills.

en you've finished, ...

the bottom of each challenge, pupils are instructed to turn to ge 80 and to find a partner or another pair or group. This page ers a structure and set of questions intended to provide pupils h an opportunity to share their results and discuss their methods, tegies and mathematical reasoning.

When you've finished, turn to page 80.

utions

ere appropriate, the solutions to the challenges in this book can be found at *Busy Ant Maths* Collins Connect and on our website: collins.co.uk/busyantmaths.

Number sequences

Solving mathematical problems

Challenge

Choose one of the numbers on the flags.

Write a sequence that has your number as its middle number.

Choose another number and do the same thing.

Now choose any two numbers and write a sequence that has your numbers in it.

Repeat, choosing other pairs of numbers.

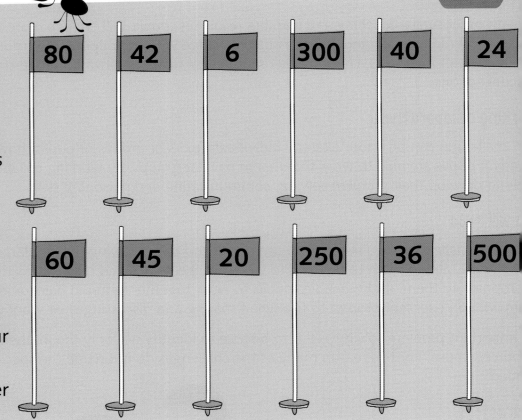

| 80 | 42 | 6 | 300 | 40 | 24 |

| 60 | 45 | 20 | 250 | 36 | 500 |

Think about ...

Think about writing sequences involving steps of 2, 3, 4, 5, 8, 10, 50 or 100.

Remember that a number sequence can count forwards or backwards.

Make sure that each sequence contains at least five numbers.

What if?

Choose any two numbers and write a sequence that has your numbers as the two end numbers.

Choose any three numbers and write a sequence that has your numbers in it.

When you've finished, turn to page 80.

Challenge

You will need:
• set of 0–9 digit cards or a 0–9 dice

Choose three digit cards or roll the dice three times.

Write down the numbers.

Use the three numbers to make as many different 2-digit and 3-digit numbers as you can.

What is 1, 10 and 100 **more** than your numbers?

Repeat, using other sets of three digit cards.

What patterns do you notice?

Think about ...

Think about which digit changes when you find 1, 10 or 100 **more** or **less**.

When does more than one digit change in a number when you find 1, 10 or 100 **more** or **less**?

What if?

What about 1, 10 and 100 **less** than your numbers?
What patterns do you notice?

When you've finished, turn to page 80.

Challenge

Using the digits 7, 4 and 8, how many different 2-digit and 3-digit numbers can you make that lie between 70 and 700?

Order the numbers, from smallest to greatest.

What other numbers can you make using the digits 7, 4 and 8?

Think about ...

Work systematically so that you can spot any patterns.

Look at the numbers that lie within the number range, and then look at the numbers that are greater or smaller than the number range.

What if?

What numbers can you make between 50 and 500 using the digits 9, 1 and 5?

What other 2-digit and 3-digit numbers can you make?

What do you notice that is similar about the numbers you made using the digits 7, 4 and 8 and those you made using the digits 9, 1 and 5?

Can you choose three other digits and a number range where the same thing happens?

When you've finished, turn to page 80.

Challenge

My number is 1 less than a multiple of 100. Its tens digit is 4 more than its hundreds digit.

All the digits in my number are even. The hundreds digit is double the tens digit. The sum of its digits is 18.

My number is a multiple of 50. The sum of its digits is 9. The hundreds digit is 1 less than the tens digit.

Each digit in my number is a different multiple of 3. The hundreds digit is half of the ones digit.

What number is each child thinking of?

Think about ...

Remember, there are 10 digits: 0, 1, 2, 3, ... 9.

For the number 149, the sum of its digits is 14 (1 + 4 + 9).

What if?

Think of four different 2-digit or 3-digit numbers. Write a set of clues for each number. Try to write as few clues as possible for each number. Your clues must lead to just one number.

Show your clues to a friend and ask them to work out what numbers you chose.

When you've finished, turn to page 80.

Challenge

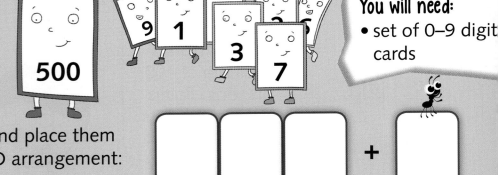

You will need:
• set of 0–9 digit cards

Shuffle a set of 0–9 digit cards.

Deal the top four cards and place them in the following HTO + O arrangement:

Add the two numbers together.

Using the same four digits, which HTO + O calculation can you make that has an answer nearest to 500?

Repeat, using four of the six remaining digit cards.

Think about ...

The answer nearest to 500 may be greater or smaller than 500.

Think about all of the possible 3-digit numbers you can make using the four digit cards.

What if?

Remove zero from the pack of digit cards and place it to one side.

Deal the top four cards and, along with the zero card, place them in the following HTO + T arrangement:

Now which calculation can you make that has an answer nearest to 500?

What if you subtracted the numbers?
Which calculations have an answer nearest to 500?

When you've finished, turn to page 80.

Challenge

Write a 3-digit number using three consecutive digits.

Reverse the digits and find the sum of the two numbers.

Repeat for other 3-digit numbers with consecutive digits.

How many different calculations can you make?

Write about any patterns you notice.

```
  1 2 3
+ 3 2 1
-------
  4 4 4
```

Think about ...

Once you've completed a few calculations, predict what you think the next answer might be.

Think about how to record your calculations, including the answers, in a way that will help you to spot any patterns.

What if?

What if you find the difference between each pair of numbers?

What do you notice?

What if you add three consecutive **even** digits?

How many different calculations can you make?

What if you add three consecutive **odd** digits?

How many different calculations can you make?

Compare your answers with the answers you got for the addition calculations in the Challenge above. What do you notice?

```
  3 2 1
- 1 2 3
-------
```

```
  2 4 6
+ 6 4 2
-------
```

```
  1 3 5
+ 5 3 1
-------
```

When you've finished, turn to page 80.

Making numbers

Challenge

Use pairs of numbers from the screen. Add the numbers or find the difference between them to make the numbers on these cards:

615 **306** **547**

245 **177** **691**

382 **752** **238**

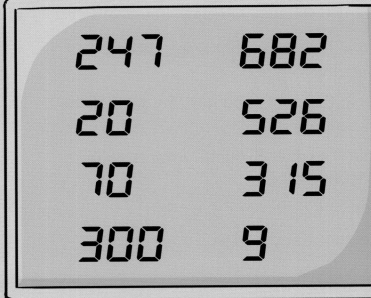

247   682

20   526

70   315

300   9

Did you use all the numbers on the screen? If not, which numbers did you not use?

Think about ...

Making estimations will help you identify possible pairs of numbers on the screen that, when added or subtracted, will be one of the numbers on the cards.

Look for similarities between the numbers on the cards and the numbers on the screen.

What if?

What other numbers can you make by adding or subtracting pairs of numbers on the screen?

When you've finished, turn to page 80.

hallenge

The Norman Conquest and the Battle of Hastings both took place in 1066.

nvestigate different addition alculations of 2-digit and 3-digit umbers that have the answer 1066.

□□□ + □□   □□□ + □□□

hink about ...

What's the easiest calculation you can make using 2-digit and 3-digit numbers? What's the hardest calculation?

Think about working backwards.

What if?

nvestigate different subtraction alculations of 2-digit and 3-digit numbers that have the answer 776.

The first Olympiad was held in 776 BC.

□□□ – □□

□□□ – □□□

When you've finished, turn to page 80.

Using tables

Challenge

Write out the multiplication facts for the 3 times table.

Investigate how you can use the 3 times table to help you with the 6 times table.

How does the 6 times table help you with the 12 times table?

$3 \times 0 = 0$
$3 \times 1 = 3$
$3 \times 2 = 6$
$3 \times 3 = 9$
$3 \times$

Think about ...

What relationship can you spot between 3, 6 and 12? What about 4 and 8?

Is there any relationship between the answers to the 3, 6 and 12 times tables? What about the answers to the 4 and 8 times tables?

What if?

How does the 4 times table help you with the 8 times table?

What about the 10 times table to help you with the 9 and 11 times tables?

How could you use the 10 and 5 times tables to help you answer 7×15?

When you've finished, turn to page 80.

Challenge

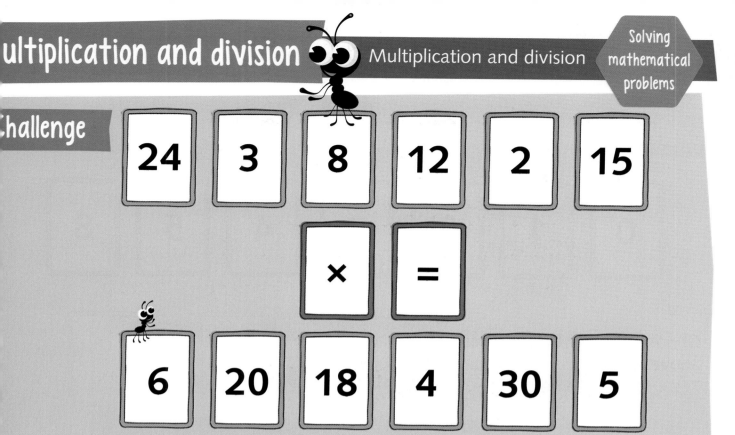

se only the numbers and signs above to make as many calculations as you can.
he answer to each calculation must also be a number on one of the cards.

Think about ...

If you can use three cards to make one calculation, how can you rearrange the three cards to make other calculations?

What relationships do you see between the different calculations you make?

What if?

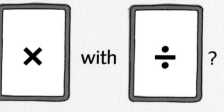

What if you replace ✗ with ÷ ?

Compare your multiplication calculations with your division calculations – what do you notice?

When you've finished, turn to page 80.

Challenge

Choose three of the digit cards above and place them in the following TO × O arrangement:

Multiply the two numbers together.

What is the greatest answer you can make?

Apart from zero, what is the smallest answer?

Think about ...

You can't use the same digit card more than once in the same TO × O arrangement. So, for example, you can't have 23 × 2 or 38 × 8.

Don't forget to 'use what you know' and use 'trial and improvement' to help you find different calculations.

What if?

Using three of the digit cards above, and the same TO × O arrangement, how many different calculations can you make that have an answer between 250 and 350?

When you've finished, turn to page 80.

hallenge

artition 16 into two numbers.

Multiply the two numbers ogether.

$9 \times 7 = 63$

artition 16 in other ways.

Vhich partitioning of 16 into wo numbers will give you the reatest product?

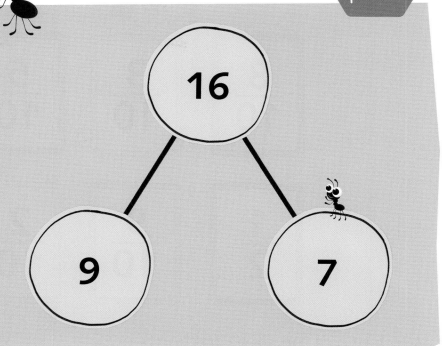

hink about ...

Think about all the different ways you can partition 16, 19 and 23.

Think carefully about how you organise your different partitionings and calculations so that you can spot any patterns that will help with your predictions.

What if?

What if you partition 19 into two numbers? Which partition creates the greatest product?

Look at the two calculations that give the greatest products when you partition 16 and 19 into two numbers. What do you notice?

Predict, and then find out, which partition of 23 will create the greatest product.

Choose a 2-digit number. Predict, and then find out, what is the best partition to make the greatest product.

When you've finished, turn to page 80.

Fractions that total 1

Challenge

$\frac{8}{10}$	$\frac{3}{10}$	$\frac{5}{10}$	$\frac{7}{10}$	$\frac{1}{10}$
$\frac{1}{2}$	$\frac{9}{10}$	$\frac{2}{10}$	$\frac{6}{10}$	$\frac{4}{10}$

Which two of the above fractions make a whole?

How many different ways can you find?

Which three of the above fractions make a whole?

How many different ways can you find?

Think about ...

Think carefully about how you're going to show which two or three fractions total 1.

Being systematic will help you identify all the different ways possible.

What if?

Which two of the above fractions make a half?

How many different ways can you find?

Can you think of other pairs of fractions that make a whole? What about a half?

When you've finished, turn to page 80.

uying sweets

hallenge

mily and Alexander have bought eight sweets between them.
vestigate what fraction of the eight sweets each child could have bought.

hink about ...

Being systematic will help you find all the different ways of sharing the sweets.

Think about how you can express fractions in different ways, for example $\frac{5}{10}$ and $\frac{1}{2}$.

What if?

What if Osaru, Emily and Alexander have bought 12 sweets between them? What fraction could each child have bought?

What if Kate, Osaru, Emily and Alexander have bought 10 sweets between them? What fraction could each child have bought?

When you've finished, turn to page 80.

Equivalent fractions

Challenge

We're equivalent.

$$\frac{1}{2} = \frac{3}{6}$$

Lay out four 1–9 digit cards, as above, to make an equivalent statement.

How many equivalent statements can you make using four cards from a set of 1–9 digit cards?

Think about ...

Think about unit fractions such as $\frac{1}{2}$ and $\frac{1}{3}$ and also non-unit fractions such as $\frac{2}{3}$ and $\frac{3}{4}$.

Think about how a fraction wall might help you find equivalent fractions.

What if?

What if you have two sets of 1–9 digit cards and can use the same number twice?

When you've finished, turn to page 80.

Challenge

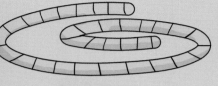

21 cm   **6 cm**   **17 cm**

12 cm   **9 cm**

Using only the five ropes above, I can measure all the lengths from 1 cm to 30 cm, except for one length.

Is Kate right?

Think about ...

You can place ropes end to end to find the total length, or side by side to find the difference.

Show how you can work out each of the different lengths.

What if?

Which five different lengths of rope would you need to make every length from 1 cm to 30 cm?

When you've finished, turn to page 80.

Challenge

Use only 50 g, 100 g and 500 g weights. Investigate how you could use these weights to balance objects with a mass of 50 g, 100 g, 150 g, 200 g … up to 500 g.

You can use the same weight more than once to make a mass, but you must use the smallest number of weights each time.

Think about ...

One way of finding a mass of 300 g is by using the scale balance with a 500 g and two 100 g weights. How could this work? How might it work for other masses?

You shouldn't use more than three weights for each mass.

What if?

Although most shops use electronic weighing scales, you might still find some shops, such as butchers, sweet shops or coffee roasters, using a Standard Metric Weight Set.

A Standard Metric Weight Set consists of the following nine weights:

| 500 g | 2 × 200 g | 100 g |
| 50 g | 2 × 20 g | 10 g | 5 g |

Can you explain why a Standard Metric Weight Set has these nine specific weights?

When you've finished, turn to page 80.

lling buckets

Challenge

Using only the four buckets on the right, how can you make the amounts from 1 litre up to 30 litres?

The buckets must e full and you must fill them s few times s possible.

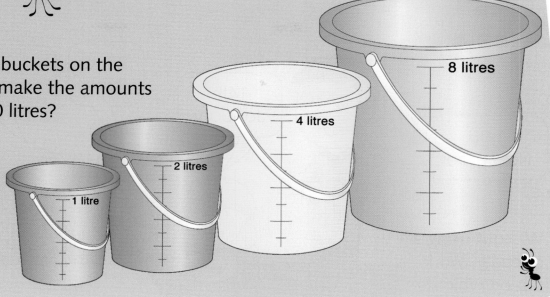

8 litres

4 litres

2 litres

1 litre

Think about ...

Work systematically through each litre, from 1 litre to 30 litres. Look out for patterns to help you.

There will be more than one way to make most amounts from 1 litre to 30 litres. Find the way that uses the smallest number of full buckets each time.

What if?

What if you have the following buckets and containers?

$\frac{1}{4}$ litre

$\frac{1}{2}$ litre

1 litre

3 litres

8 litres

How many different amounts can you measure if you can only fill each bucket or container once?

When you've finished, turn to page 80.

Challenge

14th November

14th January

22nd June

5th March

29th September

Without using a calendar, work out on which day of the week each character's birthday falls this year.

Think about ...

Perhaps start with today's date and work forwards and backwards.

Do you know of any important dates this year that occur on a particular day of the week?

What if?

This year, what date is the:

- first Saturday in October?
- third Monday in May?
- second Friday in December?
- last Sunday in July?
- first Wednesday in April?

When you've finished, turn to page 80.

Challenge

Emily has one note and three identical coins in her purse.

Investigate how much money she could have in her purse.

Think about ...

Look for a relationship between the different amounts you can make using one £5 note and the amounts you can make for each of the other different notes.

Recording your results systematically in a list will help you find all the different amounts Emily could have.

What if?

What if Emily has one coin and three identical notes in her purse?

When you've finished, turn to page 80.

Challenge

You will need:
- right angle measure
- ruler

This pentagon has two right angles.

What other pentagons can you draw that have two right angles?

Can you draw other pentagons that have more than two right angles?

Think about ...

Try to be as accurate as possible when you draw your shapes.

Mark each right angle with a small square as shown above.

What if?

Can you draw a hexagon with two right angles?
What about more than two right angles?

What about octagons?

When you've finished, turn to page 80.

Challenge

A tessellation is a pattern of shapes that fit perfectly together, with no overlaps or gaps.

You will need:
- set of 2-D shapes
- coloured pencils

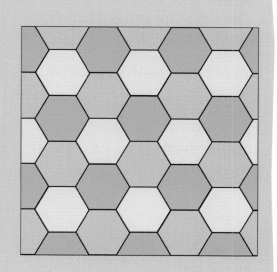

Investigate which regular 2-D shapes tessellate.

Use the shapes to make some tessellating patterns.

Think about ...

Think about using colour in your tessellations, but remember, there needs to be a pattern in your colours as well as your shapes.

Be as accurate as you can in your drawings, making sure that there are no gaps or overlaps.

What if?

Can you make a tessellation from two or more different regular 2-D shapes?

When you've finished, turn to page 80.

Brothers and sisters

Challenge

How many brothers and sisters does each of the children in your class have?

How many brothers and sisters do the parents of the children in your class have?

Draw two graphs to show your data.

Write about the similarities and differences between the data in the two graphs.

You will need:
- graph paper
- ruler

Think about ...

Think carefully about how to collect the data from everyone in the class. What is the most efficient way of doing this?

What type of graph will you use to represent your data? If you draw a bar chart, how are you going to label the axes? If you draw a pictogram, what will one picture stand for?

What if?

A family tree is a chart that shows all the people in a family over many generations and their relationship to one another.

Draw your family tree.

When you've finished, turn to page 80.

Challenge

You will need:
- graph paper
- ruler

What is your favourite television programme?

How does this compare with other children in your class?

What is the most popular television programme in your class?

Think about ...

Think carefully about how to collect the data from everyone in the class. What is the most efficient way of doing this?

How are you going to organise and represent your results to show what conclusions you make?

What if?

For one week, keep a record of how much television you watch each day.

On which day of the week do you watch the most television?

Approximately what fraction is this of all the television you watch during the week?

When you've finished, turn to page 80.

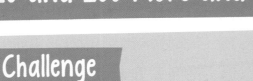
Challenge

Complete each row.

+10	+10	+10	+10	+10	+10	+10
456	466	476					

+100	+100	+100	+100	+100	+100	+100
647	747	847					

−10	−10	−10	−10	−10	−10	−10
851	841	831					

−100	−100	−100	−100	−100	−100	−100
757	657	557					

Are Kate's and Alexander's statements always true, sometimes true or never true?

Provide examples to explain why.

When finding 10 more or less than a 3-digit number, you only need to change the tens digit.

Adding or subtracting 100 to or from a 3-digit number means you only have to change the hundreds digit.

Think about ...

Look carefully at all the digits in each number.

Which digits change in the number? Which digits stay the same? Is it the same for all numbers in the patterns?

What if?

Which digits change when you find 10 more or less than a 2-digit number?

Which digits change when you find 10 more or less than a 4-digit number?

Which digits change when you find 100 more or less than a 4-digit number?

When you've finished, turn to page 80.

hallenge

What's the same about these numbers?
What's different?

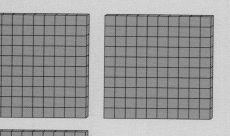

H	T	O
8	3	4

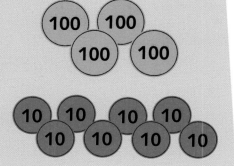

400 + 6 + 80

648

100s	10s	1s

hink about ...

Think about the value of each of the digits.

What is the difference in the value of the digit 4 in each of the representations?

What if?

738 can be written as 7 hundreds, 3 tens and 8 ones.
It can also be written as 73 tens and 8 ones.
What other ways can you express 738?

Choose a 3-digit number and express it in as many different ways as you can.

When you've finished, turn to page 80.

33

Challenge

Shuffle a set of 1–9 digit cards. (Don't include zero for this part of the Challenge.)

Place the cards face down in a pile.

Turn over one card at a time and place it in the ones, tens or hundreds position to make a 3-digit number.

Once a card is placed in position, it can't be moved.

Try to make the three greatest 3-digit numbers you can.

When you've finished, shuffle the cards and repeat.

Can you make three larger 3-digit numbers than before?

Write some rules for trying to make the three greatest 3-digit numbers possible.

Think about ...

Think carefully about whether a card should go in the hundreds, tens or ones place.

Which place value is the most important?

What if?

What if you use a pack of 0–9 digit cards? What are the two greatest 3-digit numbers and two greatest 2-digit numbers you can make?

When you've finished, turn to page 80.

Challenge

Which numbers match only one statement?

Explain why.

Which numbers match two of the statements?

Explain why.

Which number matches the most statements?

Explain why.

Which statements only match one of the numbers?

Numbers
24
50
300
75
15
32
250
700
25

Statements
multiple of 8
odd number
multiple of 50
more than 500
3-digit number
multiple of 100
less than 20
between 30 and 40
multiple of 4

Think about ...

Think about place value, comparing and ordering numbers, multiples, and other properties of number.

Think about how you could sort the numbers in different ways.

What if?

Write a different statement that matches only one of the numbers.

Write a different statement that matches two of the numbers.

For each statement, write another two numbers.

When you've finished, turn to page 80.

In your head or on paper? Addition and subtraction

Reasoning mathematically

Challenge

Which of these calculations would you do in your head?

Which would you use pencil and paper for?

For each mental calculation, show what you would do in your head.

For each written calculation, show what you would write down.

a) 475 + 60

b) 942 − 9

c) 734 − 268

d) 283 + 5

e) 706 − 400

f) 358 + 576

g) 838 − 70

h) 216 + 300

Think about ...

Make sure that you give the answer to each calculation and show all your working.

For the second 'What if?' question below, make sure that you label each group to show the criteria you have used to sort the eight calculations.

What if?

For which of the calculations above would you first make an estimate of the answer? How would you go about doing this?

Sort the eight calculations into different groups. For each group, write one more calculation that belongs in that group. Can you sort the eight calculations in other ways?

When you've finished, turn to page 80.

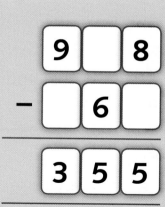

Challenge

Find the missing digits in each of these calculations.

Write about how you worked out how to find each missing digit.

$$\begin{array}{r} 4\ \square\ 7 \\ +\ \square\ 9\ \square \\ \hline 8\ 2\ 3 \end{array}$$

$$\begin{array}{r} 9\ \square\ 8 \\ -\ \square\ 6\ \square \\ \hline 3\ 5\ 5 \end{array}$$

Think about ...

Think about which operation you need to use to work out each missing digit.

Think about place value and the value of each of the digits.

What if?

Complete these calculations, using each of the digits 1 to 9 only once in each calculation.

1   2   3   4   5

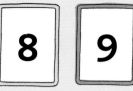

6   7   8   9

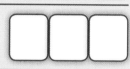

Can you write more than one addition and subtraction calculation in this way?

When you've finished, turn to page 80.

Spot the mistake

Challenge

Kate, Emily, Alexander and Osaru each made a mistake when answering a set of calculations.

Kate	Emily	Alexander	Osaru
9 5 4	3 6 1	6 2 9	7 0 4
− 4 3 7	+ 5 8 2	+ 3 5 8	− 5 2 2
5 2 3 ✗	8 1 4 3 ✗	9 7 7 ✗	2 8 2 ✗

What is the error in each of these calculations?

What is the correct answer to each calculation?

Think about ...

Think carefully about how you are going to explain the error that each child has made in their calculation.

These terms might help you explain the errors and offer some tips:

$$1\ 2\ 3 \leftarrow \text{Addend}$$
$$+\ 2\ 4\ 5 \leftarrow \text{Addend}$$
$$3\ 6\ 8 \leftarrow \text{Sum or Total}$$

$$8\ 7\ 6 \leftarrow \text{Minuend}$$
$$-\ 1\ 4\ 2 \leftarrow \text{Subtrahend}$$
$$7\ 3\ 4 \leftarrow \text{Difference}$$

What if?

What tips would you give Kate, Emily, Alexander and Osaru so that they each don't make the same mistake again?

Your tips might include an explanation and an example.

When you've finished, turn to page 80.

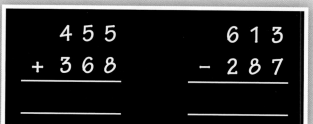

```
  4 5 5        6 1 3
+ 3 6 8      - 2 8 7
_____      _____
```

hallenge

efore Emily worked out these calculations, he estimated the answers.

I think the answer to the addition will be about 800, and the answer to the subtraction about 400.

Osaru estimates the answers to two similar calculations.

I think the answer to the addition is about 600, and the answer to the subtraction about 500.

Write an addition calculation and a subtraction calculation that Osaru could be solving.

hink about ...

Like Emily, Osaru is solving calculations involving two 3-digit numbers.

The answer to a calculation may be more or less than the estimate.

What if?

Work out the answers to the four calculations Emily and Osaru are solving.

Then, for each calculation, write a different calculation o check your answers.

When you've finished, turn to page 80.

Challenge

I know that 4 × 8 = 32. I can use this fact to help me work out the answers to these facts.

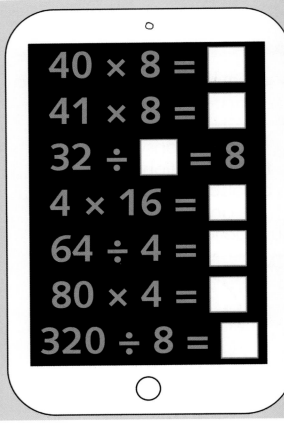

40 × 8 = ☐

41 × 8 = ☐

32 ÷ ☐ = 8

4 × 16 = ☐

64 ÷ 4 = ☐

80 × 4 = ☐

320 ÷ 8 = ☐

Explain how Kate might work out each answer.

Think about ...

Think about the relationship between multiplication and division, and what happens when you multiply a number by 10.

Write the answer to each calculation and show all your working.

What if?

Alexander says:

I know that 3 × 6 = 18

What other multiplication and division calculations might Alexander be able to work out?

Show how Alexander might work out each answer.

When you've finished, turn to page 80.

Challenge

What numbers are missing from the grey spaces in these calculations?

×		
4	240	28

```
    2 4 0
+     2 8
_____
```

×	50	
8		48

```
_____
+       4 8
_____
    4 4 8
```

Explain how you worked out the missing numbers.

Think about ...

How might the inverse relationships of multiplication and addition help you work out the missing numbers and digits?

Think about making estimates and using 'trial and improvement'.

What if?

What digits are missing from the grey boxes in these calculations?

```
  6  □
×    □
_____
  1  9  5
```

```
  □  9
×    □
_____
  3  9  2
```

```
  □  □
×    4
_____
  2  5  2
```

When you've finished, turn to page 80.

Explain how you worked out the missing digits.

Challenge

All the answers to the 4 times table are also answers to the 2 times table.

All the answers to the 8 times table are also answers to the 4 times table.

All the answers to the 8 times table are also answers to the 2 times table.

All the answers to the 2 times table are also answers to the 4 times table.

Which statements are true? Which statements are false?

Justify your decisions.

Think about ...

What are the multiples of 2, 4 and 8? What about beyond the 12th multiple?

What patterns do you notice in the answers to the 2, 4 and 8 times tables?

What if?

Emily also says:

True or false?

Prove your answer.

All the answers to the 2, 4 and 8 times tables are even numbers.

When you've finished, turn to page 80.

Challenge

Write down a multiplication act for the 2, 3, 4 or 5 multiplication ables.

$4 \times 2 = 8$

Double the first number and make a new calculation, working out the answer.

Keep doing this for as long as you can.

What do you notice about each of your answers?

Does this always work?

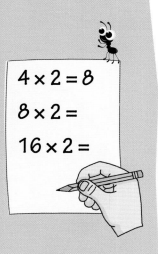

$4 \times 2 = 8$
$8 \times 2 =$
$16 \times 2 =$

Think about ...

Start with a simple multiplication fact, such as 2×3, 3×4, 5×2.

Try different starting multiplication facts.

What if?

What if you double the second number?

Compare your answers with those above.

What do you notice?

$4 \times 2 = 8$
$4 \times 4 =$
$4 \times 8 =$

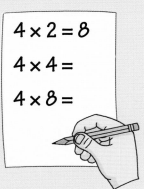

What if you double both numbers?

$4 \times 2 = 8$
$8 \times 4 =$
$16 \times 8 =$

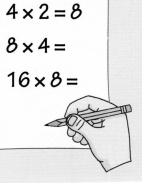

When you've finished, turn to page 80.

43

Challenge

9 is $\frac{3}{4}$ of 12.

Is Osaru right?

Prove it by drawing a picture or diagram.

Find other pairs of whole numbers where one number is $\frac{3}{4}$ of the other.

What do you notice about the pairs of numbers?

Think about ...

What patterns do you notice in the numbers? Can you use this to help you find other pairs of numbers?

How does your knowledge of multiples help you?

What if?

Find pairs of numbers where one number is $\frac{2}{3}$ of the other.

What about where one number is $\frac{3}{5}$ of the other?

What about $\frac{4}{5}$?

What about $\frac{7}{10}$?

When you've finished, turn to page 80.

Challenge

What numerators and denominators are covered by the ink splats?

Explain how you know.

$$\frac{\blacksquare}{7} + \frac{1}{7} = \frac{6}{7}$$

$$\frac{\blacksquare}{6} - \frac{2}{6} = \frac{3}{6}$$

$$\frac{2}{\blacksquare} + \frac{2}{5} = \frac{4}{5}$$

$$\frac{\blacksquare}{8} - \frac{1}{\blacksquare} = \frac{5}{8}$$

$$\frac{4}{\blacksquare} + \frac{3}{10} = \frac{\blacksquare}{10}$$

$$\frac{7}{9} - \frac{\blacksquare}{9} = \frac{2}{\blacksquare}$$

Think about ...

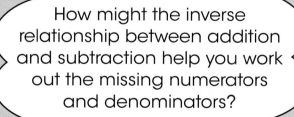

Work systematically for the 'What if?' question.

How might the inverse relationship between addition and subtraction help you work out the missing numerators and denominators?

What if?

What numerators are covered by these ink splats?

$$\frac{\blacksquare}{9} + \frac{\blacksquare}{9} = 1$$

$$1 - \frac{\blacksquare}{7} = \frac{\blacksquare}{7}$$

There is more than one solution to each calculation.

How many ways can you find?

Explain how you know you have found them all.

When you've finished, turn to page 80.

45

Challenge

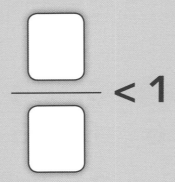

$$\frac{\boxed{}}{\boxed{}} < 1$$

Write a fraction less than 1.

Write a different fraction less than 1.

Altogether, write eight different fractions less than 1.

Which fractions are less than $\frac{1}{2}$? Which fractions are more than $\frac{1}{2}$?

Order your fractions on a number line to prove your answers.

Think about ...

Think about writing both unit and non-unit fractions.

Try to be as accurate as you can when placing your fractions on the number line.

What if?

Look at all the fractions on your number line.

Can you write an equivalent fraction for each of these fractions?

Choose one pair of equivalent fractions and draw a picture or diagram to show that they are equivalent.

When you've finished, turn to page 80.

Challenge

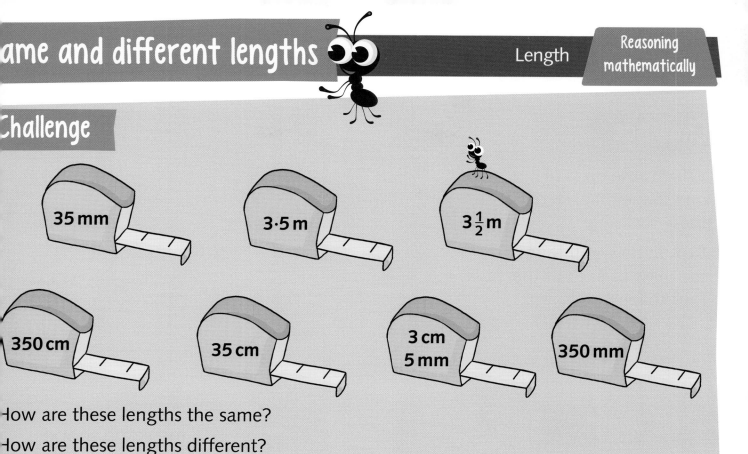

How are these lengths the same?
How are these lengths different?

Think about ...

Think about ways you can express the same measurement using different units of measure.

Use what you know about the number of centimetres in 1 metre, and the number of millimetres in 1 centimetre.

What if?

Order these measures, starting with the smallest.

300 mm	5 cm	4·5 cm	40 mm

Then write a measure that lies between pairs of measures.

When you've finished, turn to page 80.

Balance problems

Challenge

What is the mass of one shape on each of these scales?

Explain how you found the mass of each of the four shapes.

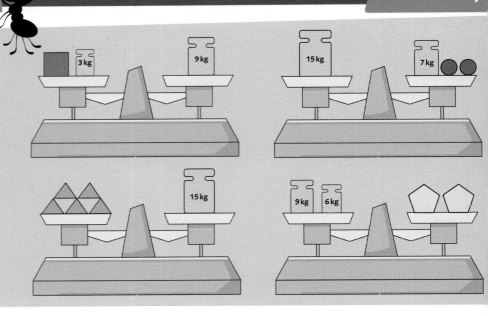

Think about ...

Remember, the scale only stays balanced when the shapes and weights on either side are equal.

If the same shape appears on both sides of a scale, you can remove a shape from both sides and the scale will stay balanced.

What if?

What is the mass of one shape on each of these scales? Think carefully!

Explain how you found the mass of each of these four shapes.

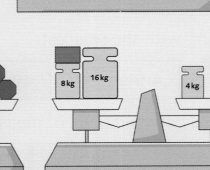

When you've finished, turn to page 80.

48

hallenge

    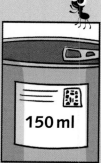

   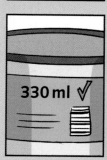 

0·5 litre

300 ml

$1\frac{1}{2}$ litres

$\frac{1}{4}$ litre

150 ml

2 litres

1000 ml

$\frac{3}{4}$ litre

330 ml ✓

1·75 litres

ut these volumes in order, starting with the greatest.

xplain your thinking.

hink about ...

Think about the different ways that you can express each of the volumes.

Compare two and three volumes using the < and > signs.

What if?

Emily wrote:

$\frac{1}{4}$ litre < 330 ml

What other true statements could you write using the < and > signs and the volumes above?

When you've finished, turn to page 80.

All the same time?

Challenge

These five clocks all show the same time.

Not necessarily.

Who's right? Explain why

Think about ...

What time of the day do these clocks show?

Think about what you're going to include on each of the clocks for the 'What if?' question.

What if?

How would you show 10 minutes to 7 on each of the five clocks, making sure that each clock shows exactly the same time of day?

When you've finished, turn to page 80.

Challenge

Cate, Emily, Alexander and Osaru all visit a shop at the seaside.

What items did each child buy?

How do you know what each child bought?

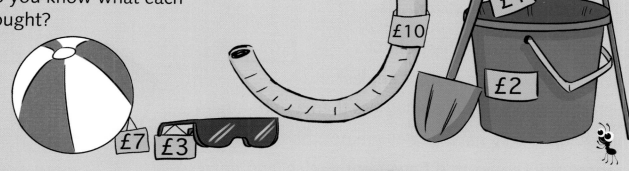

I bought 2 things and spent £13.

I spent £12 on 4 items.

I spent £23 on 3 things.

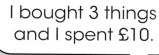

I bought 3 things and I spent £10.

hink about ...

Write a short statement explaining how you know what each child spent their money on.

What if?

Could any of the children still have bought the **same number** of items, and spent the **same amount** of money, but have bought **different items**?

Could any of the children have bought a **different number** of items but still have spent the **same amount** of money?

When you've finished, turn to page 80.

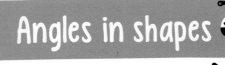

Challenge

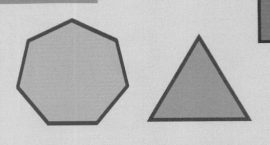

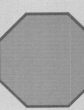

Describe the angles in these shapes.

Which shapes only have right angles?

Which shapes only have angles less than a right angle?

Which shapes only have angles greater than a right angle?

Are there any shapes that have right angles, and angles less or greater than a right angle?

Are there any shapes that have no angles?

Think about ...

Remember, **regular** shapes have all sides the same length and all angles the same size.

Remember, a **polygon** is a 2-D shape with three or more straight sides.

What if?

Emily says:

In a polygon, the number of its sides is always the same as the number of its angles.

Alexander says:

The angles in the two green shapes must be the same as both shapes are triangle

Are Emily's and Alexander's statements true or false?

When you've finished, turn to page 80.

Challenge

Which capital letters have only perpendicular lines?

Which capital letters have only parallel lines?

Which capital letters have both perpendicular and parallel lines?

Which capital letters have neither perpendicular nor parallel lines?

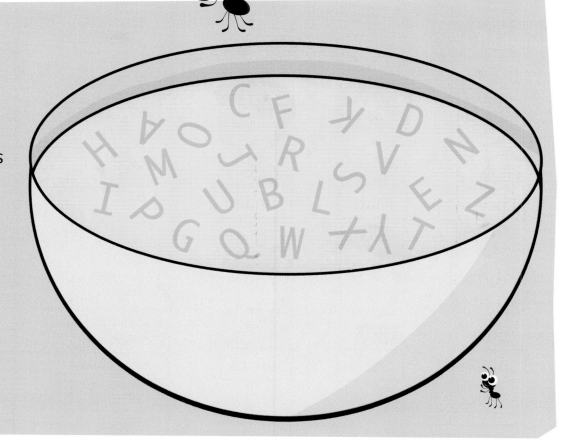

Think about ...

How might different handwriting styles and typewritten fonts affect your answers?

Think about how you're going to record your results. They should show why you have decided a particular letter does or does not have perpendicular and parallel lines.

What if?

Find someone who has got some different answers from you. Try to convince them that your answers are right.

Then work together and do the same for the 26 lower-case letters of the alphabet.

When you've finished, turn to page 80.

Challenge

Lucy

Matt

Vihaan

Pari

Sophie

Oscar

Jessica

Who is represented by each letter in this table?

Explain your reasoning.

	Has brown eyes	Does not have brown eyes
Is less than 20 years old	E	A D
Is more than 20 years old	B G	C F

Think about ...

Is there more than one possible answer for each table?

What are the strengths and weaknesses of presenting this information in tables such as these?

What if?

Who is represented by each letter in these tables?

Explain your reasoning.

	Female		Male
Is less than 20 years old	K		I L
Is more than 20 years old	H J N		M

	Has brown eyes	Does not have brown eyes
Female	S	P R U
Male	Q T	O

When you've finished, turn to page 80.

hallenge

What could this bar hart be showing?

What could the axes epresent?

What conclusions can ou draw from the ata in the chart?

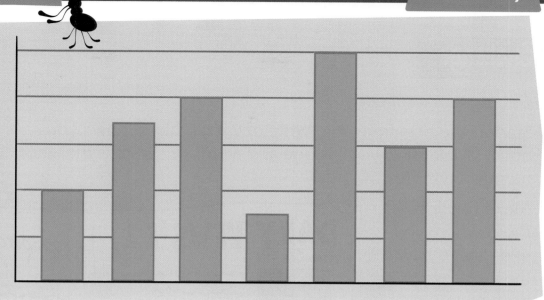

hink about ...

What if these charts were about the days of the week or the different years in a school?

What other topics might each of these charts be about?

What if?

What could this pictogram be showing?

What might the headings be?

What conclusions can you draw?

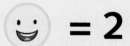 = 2

When you've finished, turn to page 80.

Numbers in newspapers

Challenge

Choose a newspaper.

Start at the beginning and write down all the numbers you notice. Make sure you include any words or symbols that the numbers refer to.

Make two lists – those where you understand how the numbers have been used, and those where you don't.

Choose two of the things from the list that you don't understand and find out what they mean.

You will need:
- selection of different newspapers

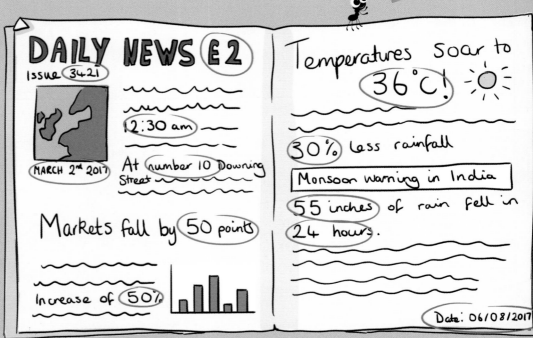

DAILY NEWS £2

Issue 3421

12:30 am

MARCH 2nd 2017

At number 10 Downing Street

Markets fall by 50 points

Increase of 50%

Temperatures soar to 36°C!

30% less rainfall

Monsoon warning in India

55 inches of rain fell in 24 hours.

Date: 06/08/2017

Think about ...

Look for examples of numbers in different contexts.

What are most of the numbers in the newspaper used to describe?

What if?

Look for examples of different charts, graphs and tables in the newspaper.

When you've finished, turn to page 80.

Binary numbers

Challenge

A binary number is made up of only 0s and 1s. The digits 2, 3, 4, 5, 6, 7, 8 and 9 are not used in binary numbers.

Look at the binary numbers for the Hindu-Arabic numbers 1 to 10.

What patterns do you notice?

Hindu-Arabic number	Binary number
1	1
2	10
3	11
4	100
5	101
6	110
7	111
8	1000
9	1001
10	1010

You can work out binary number equivalents for numbers in the Hindu-Arabic number system using this number converter.

512	256	128	64	32	16	8	4	2	1	Hindu-Arabic number
									1	1
								1	0	2
								1	1	3
							1	0	0	4
							1	0	1	5
							1	1	0	6
							1	1	1	7
						1	0	0	0	8
						1	0	0	1	9
						1	0	1	0	10

Binary number

Copy the number converter above and include the Hindu-Arabic numbers 11 to 20 in the right-hand column. Use your converter to work out the binary number equivalents for the numbers 11 to 20.

Think about ...

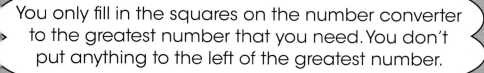

You only fill in the squares on the number converter to the greatest number that you need. You don't put anything to the left of the greatest number.

What if?

Write any ten 2-digit Hindu-Arabic numbers greater than 20 and write each number as a binary number.

Write any ten 3-digit Hindu-Arabic numbers and write each number as a binary number.

When you've finished, turn to page 80.

Challenge

The secret code to open the school safe is 2 5 8 8 7 7.

You will need:
- squared paper

1	2	3	4	5	6	
						2
						5
						8
						8
						7
						7

Design a pattern where the secret code is 4 9 3 5 4 6.

1	2	3	4	5	6	
						4
						9
						3
						5
						4
						6

Is there more than one way of making a pattern for this code?

Draw a 6-digit pattern for a friend to solve.

Can they identify the secret code?

Think about ...

What is the most difficult pattern you can make? Why is it so difficult?

When drawing your own pattern for a friend to solve, don't write the numbers down the side.

What if?

Design a code of your own.

Your code must include numbers.

Your code must be used to decipher word messages.

Write out the following message using your code:

WHEN SPIES USE BUGS AND CODES FOR SOLVING MYSTERIES, THEY JAIL VILLAINS QUICKLY.

Use your code to write out your own secret message. Give it to a friend to solve.

When you've finished, turn to page 80.

Using and applying mathematics in real-world contexts

Challenge

At the end of the Rangers v Volves match, he score was – 3.

What could the alf-time scores ave been?

Think about ...

Think carefully about any differences in the scores there may be at full-time as opposed to half-time.

Record your results systematically.

What if?

What if the full-time score was 5 – 4?

What about a score of 2 – 1 at full-time?

Look carefully at your results for each of the three matches. Make a rule for working out how many different half-time scores there are using just the final score.

When you've finished, turn to page 80.

Dice totals

Challenge

Roll two 1–6 dice.

Count the total number of dots on the top of both dice.

Write down the total number.

Do this 20 times.

What is the most common total? Why do you think this is?

What is the least common total? Why do you think this is?

Predict, then test, what you think would be the most and least common totals using two 1–10 or 0–9 dice.

Explain your prediction and results.

You will need:
- three 1–6 dice
- two 1–10 or 0–9 dice
- two 1–12 dice

Think about ...

How are you going to keep a record of your 20 rolls of the dice and your totals?

Use your results for two 1–6 dice to help predict your results for other combinations of dice.

What if?

Predict, then test, what you think will be the most and least common totals using two 1–12 dice.

Predict, then test, what you think will be the most and least common totals using three 1–6 dice.

Explain your predictions and results.

When you've finished, turn to page 80.

ice operations

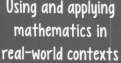

Challenge

Roll three 1–6 dice.

What is the greatest total you can get using the numbers of dots on the dice and any of the four operations?

Roll the three dice several times, each time aiming to make a larger total than before.

You will need:
- several 1–6 dice
- several 0–9 dice

Think about ...

Which operations will you use?

Will you use the numbers rolled as three separate numbers, such as 6 + 5 + 2, or will you use two of the numbers to make a 2-digit number, such as 62 + 5?

What if?

What if you use four 1–6 dice?

What if you use 0–9 dice?

When you've finished, turn to page 80.

Fractions of a newspaper

Challenge

Choose a newspaper.

What are the main sections of the newspaper?

Investigate what fraction of the whole newspaper is used for each of the different sections.

You will need:
- selection of different newspapers

Think about ...

The different sections of a newspaper might include news, business, entertainment, sports, travel …

Think about how you're going to represent your findings as fractions. You could use a diagram similar to this, which is divided into tenths:

Sections of a newspaper

What if?

Look at different pages in the newspaper.

What fraction of each page is:

- headline
- 'body' of the article
- photographs, illustrations or diagrams?

When you've finished, turn to page 80.

Tailor made

Challenge

A tailor is going to make some clothes especially for you.

Draw a simple figure of yourself.

Ask a friend to measure you and write the measurements on the figure you have drawn.

You will need:
• measuring equipment

Think about ...

What are all the different measurements the tailor will need to make your clothes?

Make your measurements as accurate as possible.

What if?

What is the mass of the clothes you wear? (Don't forget your shoes!)

How is this different at different times of the year?

How is this different when you're doing different things?

When you've finished, turn to page 80.

Walking

Challenge

You will need:
- measuring equipment

How fast do you walk?

Estimate first, then find out.

Think about ...

What will you need to use to find out how fast and how far you walk?

What units of measure are you going to use? Are you going to use a standard unit of measure such as metres, or an informal measure such as paces?

What if?

Approximately how far is your home from the nearest shop?

Estimate how long it would take you to walk from your home to the shop.

Show all your working out.

CORNER STORE

When you've finished, turn to page 80.

a minute

Challenge

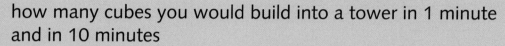

In 30 seconds, build a tower of interlocking cubes.

How many cubes are in your tower?

What is the height of your tower?

What is the mass of your tower?

Use this information to estimate:

  how many cubes you would build into a tower in 1 minute and in 10 minutes

  how tall your tower would be after 1 minute and after 10 minutes

  the mass of your tower after 1 minute and after 10 minutes.

Do you think your estimates for 1 minute will be accurate? What about for 10 minutes? Why? Why not?

You will need:
- clock or watch with a seconds hand or a stopwatch
- interlocking cubes
- water, yoghurt pot and bucket
- other readily available classroom resources
- measuring equipment

Think about ...

Measure the height, mass and volume as accurately as you can, using the appropriate units of measure.

Show how you worked out your estimates for 1 minute, 5 minutes and 10 minutes.

What if?

What if you pour yoghurt pots full of water into a bucket in 1 minute?

How many litres and millilitres of water are in the bucket?

Estimate how many litres and millilitres of water would be in the bucket after 5 minutes and after 10 minutes.

When you've finished, turn to page 80.

Water usage

Challenge

Personal usage	Average water usage
having a drink of water	$\frac{1}{4}$ litre
flushing the toilet	6 litres
washing your hands or face (with tap running)	5 litres each minute
brushing your teeth (without tap running)	2 litres
taking a shower	8 litres each minute
taking a power shower	20 litres each minute
having a bath (half-full)	80 litres
Household usage	Average water usage
washing dishes in the sink	10 litres
using a full-sized dishwasher	15 litres
using a washing machine (full load)	80 litres
cooking and preparing food	15 litres
using a watering can on the garden	5 litres a can
using a hosepipe on the garden	15 litres each minute

You will need:
- measuring equipment

Use the 'Personal usage' activities in the table to help you keep a record of how much water you use in a wee

For the 'Household usage' activities, talk to an adult to find out your share of how muc water was used during the week.

Think about ...

Think about how you might divide the amount of household water usage between everyone in your home to find out your share of the water used.

Think about how you're going to record your results.

What if?

The average person uses about 150 litres of water a day.

Is this more or less than the amoun of water that **you** use each day?

The average person uses about $\frac{6}{10}$ of their daily water usage for personal usage.

Is this the same for you?

When you've finished, turn to page 80.

orkout

hallenge

You will need:
- clock, watch or stopwatch

Design a 5-minute workout.

Write down your plan so that someone else can read it and do your 5-minute workout.

hink about ...

Think about how many different exercises you are going to have and how long each exercise will take.

What is the best way to display your workout so that it looks interesting and is easy to follow?

What if?

World Health Organization

The World Health Organization (WHO) recommends that children and young people aged 5–17 do at least 60 minutes of physical activity each day. This includes play, games, sports, walking, running and doing physical chores, at home, in school and elsewhere.

How much exercise do you do each day?

How much more or less is this than the World Health Organization's recommended one hour?

When you've finished, turn to page 80.

Youngest and oldest

Challenge

Who is the youngest child in your class?

Who is the oldest?

How old are they?

Can you work out how old they both are today?

What is the difference in their ages?

Think about ...

Think about how many years, months, weeks and days old someone is to describe how old they are.

What information do you need to know, and what resources do you need to use, to help you?

What if?

How many living relations do you have?

When was each person born?

How old is each person?

When you've finished, turn to page 80.

Challenge

7 Rue de France

Nice 06000

France

Friday 4th August 2017

Dear Aunty Joan,

Thank you for the lovely dress you gave me for my birthday last Saturday. We arrived in Nice 2 days ago. The weather is great. Yesterday and today we went to the beach. If it is sunny tomorrow mum said we can go on a picnic. A week tomorrow we are going to the mountains until the following Friday morning. The day after we arrive in the mountains my friend Sarah and her parents arrive. She is 12 years old and great fun to be with. She is 4 years older than me but her birthday is on the same day as mine. When we leave the mountains we are going back to Nice for 3 nights then we fly back home, but I don't have to go back to school for another 2 weeks after that.

Love from Kate

xx

Draw a calendar to show the date and day of the week for each of the events in the letter.

Think about ...

How many months will your calendar need to have?

Once you've drawn your calendar, re-read Kate's letter to make sure that your calendar is correct.

What if?

On what date were Kate and Sarah born?

Write about how you worked out each date.

When you've finished, turn to page 80.

School road

Using and applying mathematics in real-world contexts

Challenge

Investigate the road that your school is on.

On average, how many vehicles travel on the road each hour / each day?

When is the road busier / quieter throughout the day / week?

What fraction of the vehicles are cars / lorries / motorcycles / bicycles?

Is there any provision for people crossing the road? How long do people have to wa to cross the road? Does this differ at different times of the day? Is there a need for a pedestrian crossing or lights?

Think about ...

As you are going to have to collect data over a period of a week, think about the best way of doing this.

You can't spend all day, every day, for a week watching your road! You will have to base your conclusions on approximations. Think about how best to do this.

What if?

How does the road your school is on compare with the High Street?

What about the street where you live?

When you've finished, turn to page 80.

ublic transport

Challenge

Investigate the public transport in your local area.

Consider:

the different types of public transport

where you can travel to

how often each type of transport runs

if there is enough of each type of transport

how punctual each type of transport is

how much each type of transport costs.

Write a report about public transport in your local area.

You will need:
- public transport timetables

Think about ...

You can find most public transport timetables on the internet.

Include facts and statistics in your report. Think about including numbers in different contexts, such as frequency (how often something happens), times and prices.

What if?

Plan a trip by public transport to a place you would like to visit.

Consider:

- how you will get there, by train, bus, coach ...
- when you are going to travel, what time of the day and which day of the week. Will this make a difference to the cost?
- who you will take with you. Don't forget to allow for their fares as well. Will their fare be the same as yours?

When you've finished, turn to page 80.

Clothes 4-U

Challenge

Look through the pages of a clothing catalogue.

Choose a complete new outfit for yourself.

How much will this cost you?

You will need:
• clothing catalogu
  (printed or onlin

Think about ...

For each item of clothing, think about size, colour and price.

Don't forget a new pair of shoes!

What if?

Imagine you are going away on holiday for a week.

Where are you going?

Choose the clothes you want to take with you.

Make sure they're appropriate for where you're going and the time of year.

How much will this cost altogether?

When you've finished, turn to page 80.

School perimeter

Challenge

You will need:
- squared paper
- ruler
- other measuring equipment

Draw a plan of your school's perimeter.

Try to draw your plan to scale.

Think about ...

If your school has a fence around it, then this is probably your school's perimeter. If your school doesn't have a fence around it, then think about what you are going to consider as the school's perimeter.

What measurements are you going to need to take, and how will you go about taking these?

What if?

Can you work out the approximate perimeter of your school?

When you've finished, turn to page 80.

73

Investigating tangrams

Challenge

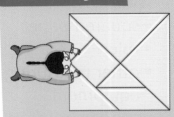

The tangram was invented in China more than 4000 years ago. According to legend, a man called Tan was taking a ceramic tile to the Emperor when he slipped and dropped it. The tile broke into seven pieces. While he was trying to put the tile back together, he found he could make many different figures and designs.

You will need:
- paper or card
- ruler
- scissors

Follow these steps to make a tangram.

1. Draw a large square with sides of at least 10 cm.

2. Mark the half-way points A, B, C and D.

3. Mark the quarter-way point E.

4. Use these as guides to draw the lines on the square.

5. Cut out the seven pieces of the tangram.

Use your tangram pieces to make different shapes and objects.

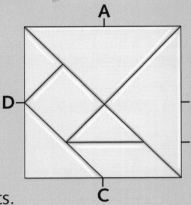

Think about ...

Make sure that you are as accurate as possible when drawing and cutting out your tangram.

Draw the different shapes and objects you make, showing how you used each of the tangram pieces.

What if?

Look at the seven shapes that make up a tangram.

Write about the different shapes and sizes.

What relationships can you see between the different shapes and sizes?

Can you express each of the tangram pieces as a fraction of the whole square?

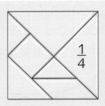

When you've finished, turn to page 80.

Tangram animals

Challenge

Use all the tangram pieces to make these animals.

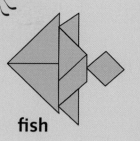

fish

horse

You will need:
- tangram
- paper or card
- ruler
- scissors

Which of these animals can you make using all seven tangram pieces?

swan

rabbit

cat

dog

What other animals and objects can you make using some or all of the seven tangram pieces?

Think about ...

Draw the different animals and objects you make, showing how you used each of the tangram pieces.

When designing your own tangram for the 'What if?' question think about:

- the shape you are going to make your tangram
- the different shapes that will make up your tangram
- the number of pieces your tangram will have.

What if?

Design your own tangram.

What shapes, animals and objects can you make?

How does your tangram compare with an ordinary seven-piece tangram? Is it as good? Why? Why not?

When you've finished, turn to page 80.

Mazes

Challenge

Write directions for moving through the maze.

Think about ...

What do you need to do before you can start to write a set of directions for moving through the maze?

Use terms such as: **quarter turn, right angle, right, left, forwards, clockwise** and **anticlockwise**.

What if?

Now do the same with this maze.

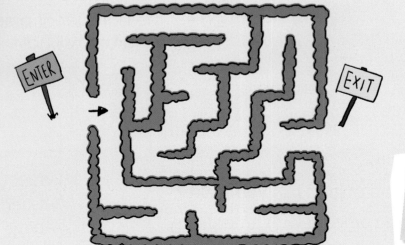

When you've finished, turn to page 80.

ight-angle programmes

Challenge

You will need:
• television guide

Look at a television guide.

Choose a channel.

Investigate how many right angles the minute hand turns through the different programmes during the day.

Think about ...

If a programme starts at 25 past 6 and finishes at $\frac{1}{4}$ past 7, that's three complete right angles:

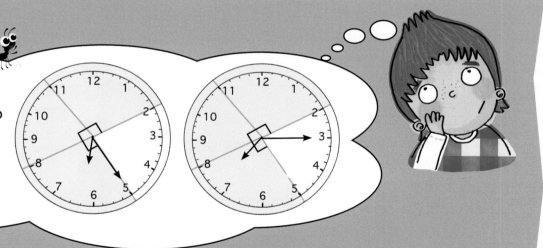

What if?

Look at a selection of TV channels on one day. During which programme or programmes does the minute hand turn through the most right angles?

What about the fewest right angles?

When you've finished, turn to page 80.

77

Home-school directions

Challenge

Imagine you have a cousin who lives in Australia.

He is coming to visit you with his family for the first time.

When your cousin arrives at your home, you want him to come straight to school to meet you.

Write an email to your cousin, giving directions from your home to your school.

Draw a simple map to go with your instructions.

You will need:
- squared paper
- coloured penci

Think about ...

Think carefully about the best way to present your set of instructions so that it is easy for your cousin to read and follow.

Can you also give some guidance on the approximate distance of each part of the journey?

Include in your directions how long the journey should take.

What if?

What if your aunt and uncle email you back and ask for directions from the airport to your home?

When you've finished, turn to page 80.

Animal classification

Challenge

Animals can be classified in different ways.

One method of classification is shown on the right.

Ask ten friends to each name five different animals.

Once you've collected the data, sort all 50 animals into the six different animal classifications.

When you've sorted your data, present your results in a bar chart.

You will need:
- squared paper
- ruler

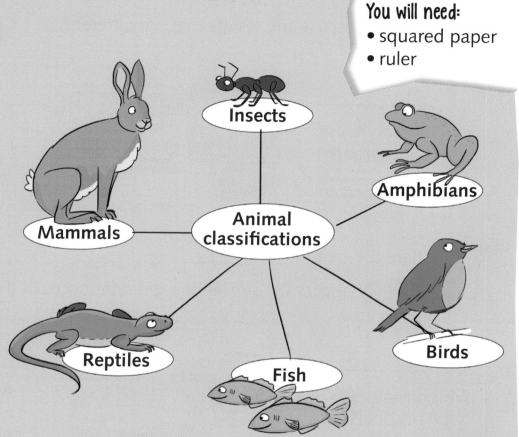

Insects

Amphibians

Mammals

Animal classifications

Birds

Reptiles

Fish

Think about ...

Think carefully about how you're going to collect, record and organise your data.

How are you going to label the axes on your bar chart? For the 'What if?' question, what symbol are you going to use for your pictogram and what will one symbol represent?

If you're not sure what classification an animal belongs to, how are you going to find out?

What if?

What if you present your results in a pictogram?

When you've finished, turn to page 80.

Share Share your results.

Discuss Discuss any results that are different.

Which result is correct?

Might there be more than one solution?

Share Share the methods used.

Discuss Discuss the similarities and differences in the methods used.

Which method worked best?

Are there any other ways to go about solving the problem?

Share Share what you have learned.

Discuss Discuss what you would do the same, and what you would do differently next time.

Is there anything you would do differently?

What have you learned for next time?

What would you do the same?